How To Grow Healthy & Tasty Vegetables

2 books in 1

Disclaimer

This book is designed to provide condensed information. It is not intended to reprint all the information that is otherwise available, but instead to complement, amplify and supplement other texts. You are urged to read all the available material, learn as much as possible and tailor the information to your individual needs.

Every effort has been made to make this book as complete and as accurate as possible. However, there may be mistakes, both typographical and in content. Therefore, this text should be used only as a general guide and not as the ultimate source of information. The purpose of this book is to educate.

The author or the publisher shall have neither liability nor responsibility to any person or entity regarding any loss or damage caused, or alleged to have been caused, directly or indirectly, by the information contained in this book.

Table of Contents

How To Grow Healthy & Tasty Cucumbers

Growing a plant, any kind of plant for that matter, is an art in itself. Growing cucumbers is especially so. It not only is tasking to attempt to grow healthy cucumbers that are tasty and fast selling; it is a lot of fun as well. This book aims to help you combine both functionality and the element of fun as you go about growing your cucumbers. This book aims to provide you with the right information to grow cucumbers that are healthy and tasty, given whatever tools you have at hand.

This is a direct book by all means, structured in a way that helps you grasp tips and useful bits while taking as little time as possible. Since it is written with the intention to aid all sorts of persons, it will also simplify the processes and elements to input in a basic, pared down way.

The cucumber, while not recognized as such, could well be the nation's plant of pride, just as the clover is to the Celts. Read on to pick up the finest tips available to help you own the most attractive cucumber garden around.

Growing Cucumbers: Basics To Know When Starting Out & Soil, Planting And Care

"The beginnings are everything; take them seriously"

- Hugh Grant

Part I: Growing Cucumbers: Basics to Know when starting out

The first thing you need to know about cucumbers is that they are tropical plants. They tend to do best when temperatures are high and water is in plenty. This being said, growing cucumbers is meant for the warmer weather. Cucumber plants are so tender, with regard to frost, that you must not set them in the garden until the soil temperatures have reached the reliable range (that is the 70 degrees range). Even when you have ascertained the soil temperatures to be safely in the 70s, absolutely desist from setting your plants in the soil until two weeks after the last recorded frost date.

Your cucumber plants will take on two growth forms: bush and vining. The vines will grow along the ground or climb up trellises if they are available. The bush kind (such as the bush hybrids) will generally take on a more compact, space-friendly form. You must understand that in general, the vining cucumbers tend to have a higher degree of produce, compared to the bush kind. However if you are strapped for space, or are into indoor gardening, the bush kind will suit your space needs best as they are better suited. You may boost the overall yield of the season (strictly with regard to the bush cucumber kind) by planting a number of cucumber crops in close succession (about two weeks apart).

Whether your cucumber planting is geared towards slicing or simply picking, you are assured of getting enough variety to make your pick from. The lemon cucumber will give you smaller fruits that are just perfect for a singular serving while the Boston picking kind has a classic heirloom taste to it that is just phenomenal. The long Armenian cucumber is an ethnic cucumber that has been pried for years for its great taste as well as the fact that one cucumber provides a numerous amount of slices. With cucumber choices, it is impossible not to find the right kind for you.

Part II: Soil, cucumber planting and overall care

Of course, cucumbers will need fertile soils if they are to grow healthy as well as tasty. But the soils need to be warm too, as the subchapter above stresses. The pH needs to be slightly acidic (in the region of 6.5). However, there are allowances for acidity as high as 6.0 or as low as 6.8. Going past these however is taking unnecessary risks.

Tips on what to do when planting your cucumber plants

Planting basics

The first step is composting. You need to work some composted manure into your soil (for fertility purposes).

Once this is done, get good quality seedlings. Take care to plant your seedlings 36-60 inches apart. This will be fully reliant on the variety you have. It is equally helpful to refer to the tag. There will always be one.

For the vine kind that is trained on trellises, space your plants about 1 foot apart.

If you are in an area where spring season is both cool and long, you should warm your soil to about 4 degrees by covering up your hill or row with some black plastic. If you prefer to keep off black plastic in your planting endeavors, you should mulch with any pine straw, chopped leaves or wheat straw, just to name a few.

Basics on taking care of your young cucumbers

If the weather is cool to the point where you consider it to be unseasonably cool, you really ought to wait until you are sure the soil has been sufficiently warmed by the sun.

Mulch will be very important if you are to keep your cucumbers clean (usually in the case of bush type cucumbers or those vining kinds that you are not training on a trellis).

Straw mulch is immensely uncomfortable for slugs as well as providing some very unsure footing for cucumber beetles. If you can access straw mulch, it should be your first priority.

If it is possible to accomplish, take time to trellis your vines. This will not only save you a lot of space, it will also aid you in your endeavors to keep your cucumbers clean. Here are some specific measurements you may keep as standards: a 15 inch diameter cage that has been fashioned from 4.5 foot (or so) welded fencing wire or even hog wire can support up to 3 vines. Wire makes the climbing efforts of the cucumber tendrils as the plant grows easier.

Cucumbers grow fast and at the same time, do not demand a lot of care. All you need to do is keep your soil consistently moist with about an inch of water weekly. If the temperatures are really high, increase that amount of water. Do the same if rains are especially scarce.

If it is possible, you should opt to use a soaker hose for your watering efforts. Drip irrigation also produces desirable results. This kind of watering also helps keep leaf diseases at bay.

You may fertilize your cucumbers with liquid food. A good example is the Bonnie range of liquid plant food. Use the liquid food about twice every month (you may do your applications after every two weeks for consistency). Apply the liquid food directly in the soil that is around the stems of the plant.

Granular, slow-release fertilizer is another option. Work it into the soil after you do your planting or simply sprinkle it around the plants later.

*Key point/action step
If you want the maximum yield for your cucumber garden, you must set up a strong foundation. The foundation here means ensuring that you provide the right environmental conditions and spacing the cucumbers properly.

Step By Step Actions To Take For Healthy & Tasty Cucumbers: Comprehensive Steps For Planting, Taking Care & Dealing With Pests

"Without the trunk, how can you have anything? Only a fool ignores the middle parts, regardless of how unspectacular they seem."

-Anon

Cucumbers are veggies that thrive best in warm weather. Before you delve into the finer bits of planting and taking full care of your cucumber patch, you must keep in mind to plant them at least 2 weeks after the last frost date, lest they fall victim to the harmful effects of frost. Especially being that they are immensely frost sensitive, it is particularly important to keep this in mind.

Most cucumber varieties are not choosy- they will happily grow in any space, especially the vining kind (attributable to its ability to climb). The most common slicing cucumber varieties will have sprawling vines that have large-sized, green leaves and tendrils that curl. Cucumbers grow at quite some speed and if you follow the directives that this chapter provides, your returns will be consistently abundant.

Planting your cucumbers

Cucumbers must be introduced into the soil no earlier than two weeks after the last recorded frost date. This is owing to the fact that cucumbers are very sensitive to frost and as such, are very susceptible to the damage that comes along with frost. Your soils must be at least 65 degrees Fahrenheit before you do any planting, though the recommendation is to have soil temperatures in the 70s.

In order to have an early crop, start your seeds indoors some three weeks before you go about transplanting them in the soil. Cucumbers prefer a bottom heat of around 70 degrees Fahrenheit (around 21 degrees Celsius). If at all, a heat mat is not available, place your seeds flat on top of your refrigerator, or place a few of them on top of your water heater. It will go a long way in helping out.

Before you go about planting, select a spot for planting. Basically, an ideal spot is the site that gets a generous amount of sun and preferably, is shade-free.

Ideally, the soils should be neutral, slightly acidic or slightly alkaline. You can improve your clay soil by adding in composted manure (to boost the fertility). If you only have dense soils that are heavy, you can improve them by adding peat to them. Rotted manure or compost will also do an equally good job. If you are unsure about the kind of soil that you have to deal with, it will do you good to have a soil test done. In the case of Northern gardens, light, sandy soils are preferred. This is because they tend to warm more quickly, compared to the other soils.

Mix in some compost or aged manure in your soil then plant about 2 inches deep. Proceed to work into the soil about 6 to eight inches in depth. Be sure to make sure that your soil is well drained and moist. Soggy soils will only reduce your chances of growing some fine, healthy and tasty cucumbers.

Sow your seeds in neat rows. Let the depth be roughly 1 inch deep and the spacing be 10 inches apart.

If at all you are transplanting your seedlings, you will need to be a bit more liberal with your spacing. Allow a 12-inch space between individual seedlings.

If you want your vine to climb (which is what you should want), a trellis is a phenomenal idea. Have one around for this purpose. The trellis will save you a lot of growth space. In addition, a trellis will protect your cucumbers from damage caused by the cucumber fruit lying on moist ground.

Caring for your cucumbers

When you are planting your seeds into the ground, take care to cover with some netting material. As a substitute, you may use a berry basket. The purpose of this is to keep off pests from digging out the seeds to feed on.

Immediately your cucumber seedlings emerge, begin to water them. Keep your watering sessions frequent. After you observe the formation of fruit, increase the amount of water you use for water to about a gallon a week.

Observe the growth of your seedlings keenly. Once they have reached 4 inches in height, thin them. You will now observe a one and a half feet space between individual plants. This is what you should aim for.

If you took the pains to add in organic matter to your soil before you set about planting your cucumbers, you may only require to side dress your plants with some well-rotted manure or some compost. If you so wish, you can opt for a fertilizer from your local store. Select one that is low in Nitrogen levels but high in potassium and phosphorous. Apply this at planting, exactly one week after blooming and every three weeks with some quality liquid food. Target the soil that is around the plant and apply directly to it. Over-fertilizing will only get your plants stunted, so avoid it.

Water your plants consistently. You are not going to raise healthy cucumbers that are also tasty if you do not water your plants religiously. If anything, chances are you will not get to enjoy the fruits of your efforts if you do not take watering seriously. Stick your finger into the soil. If you notice dryness past the first joint of your finger, then by all means understand that your plants need to be watered. You should also know that inconsistent watering will only lead to fruit that has a bitter taste to it. Let your watering sessions be especially slow: water in the morning and afternoon. Always take care to keep the watering hose away from the plant leaves.

You should mulch. Mulching is beneficial in that it helps keep in the soil moisture.

In the case of limited space or you simply want vines that are vertical, set up trellises early. This is to avoid any damage to the seedlings as well as the vines.

Take time to spray your vines with sugar water. This will attract bees and allow the setting of a higher number of fruit.

Fruiting problems & dealing with pests

To begin with, understand that your cucumbers may not set fruit if at all the first flowers were all male flowers. You should know that both male and female flowers must be growing at the same period. This may well not happen early in your plants' life. So what is the solution? Simply be patient and sit tight. Eventually, the gender balance will naturally establish itself and you will be able to enjoy observing your first fruit emerge.

The lack of fruit may also be attributed to poor pollination by bees. This is especially so if the bees are prevented from effective pollination by factors such as insecticides, cold temperatures and rain. You must keep in mind that gynoecious hybrids must have pollinator plants.

Dealing with cucumber beetles

What are cucumber beetles? If you discover that the stems of your seedlings are being bitten off, your cucumber leaves are yellowing and wilting and there is the presence of holes, then you may have stripped cucumber beetles to deal with.

Often, these beetles leave their sites of hibernation early on in the season to feed on your seedlings as they emerge (they are strongly attracted by cucurbit veggies. Cucumbers are, of course, a member of these. Others include pumpkins, squash and beans). Their attacks on your seedlings will often end up in your plants dying early on. This is especially so when the ravages of the cucumber beetles combine with the destructive action of the beetle larvae on the roots of your seedlings.

Controlling cucumber beetles

Inspect your newly planted cucumbers for the presence of this beetle. Especially when your plants are still seedlings, be very watchful.

Cover your seedlings with row covers. However, you must remember to remove these for a few hours every day during blossoming time to allow for pollination.

Tilling your garden in late fall will expose any cucumber beetles hiding there to the harsh conditions of winter and thus, cut down on their population the next year. Tilling will also make your soils easier to work on in spring.

Use natural predators for these beetles. A good example is soldier beetles. You could also opt for braconid wasps as well as some nematodes.

Dealing with white flies

Spray your cucumbers with some choice insecticidal soap. Follow up this process a couple of times, or even three.

Cucumbers are especially sensitive to insecticides. If you can introduce spiders and ladybugs among your plants, this will be especially good to combat white flies. They serve as a very potent control for these flies.

Try this mixture to put off whiteflies and control them: in a spray bottle, mix some rubbing alcohol (two parts of it) with 5 parts of water and a tablespoon of liquid soap. Spray this mixture on the cucumber foliage, targeting those that you suspect to be under the ravages of whiteflies.

*Key point/action step

You need to put in a lot of effort to grow healthy and tasty cucumbers. You need to put your focus on mulching, fighting white flies, and cucumber beetles if you really want to have the best yield. You should as well take care to maintain the right pH as well as provide all essential nutrients for maximum growth.

Bush Cucumbers Versus Vining Cucumbers: What You Need To Know About Growing Both Kinds So As To Know Which To Choose For The Best Quality Fruits

"There are no ghosts, save for them that come back to remind you of some poor choice. So choose wisely...every time"

- Soto

Gardeners often have many choices, especially when it comes to veggies. With cucumbers, this is no different. Actually, even that cucumber that an adept farmer or consumer may consider lowly comes in hundreds upon hundreds of unique varieties that have been bred for use either as slicers or as picklers. With the modern cucumber plants, growth is often conducted to include as few male flowers as possible. This has the effect of increasing the produce. Little wonder then, that some individual would have hooked his attention onto the bush kind cucumber and endorsed it as a great garden plant even when the vining cucumber was still in widespread use.

Here is what you need to know about both kinds before you make up your choice on; your cucumber of choice.

The bush cucumber

Primarily, these are bred for their phenomenal space saving qualities, seeing as they use up very little space with their very short vines. Most varieties will only call for a maximum of 3 square feet per plant.

The cultural requirements are similar to those of the vining cucumber and they mature and ripen at just about the same time and rate. Examples of popular bush cucumber varieties include the bush champion, pickle bush, parks bush whopper, salad bush, potluck and space-master.

The advantages of bush cucumbers

If you are a fan of indoor gardening, then the bush cucumber is what you should opt for. It is ideal for indoor container gardening as well as small gardens.

Bush cucumbers are not very fussy with where you plant them, though you must ensure that you provide optimum growth conditions for your plant. However, they will grow phenomenally in just about any well-drained space you plant them in, in your garden. Especially if the air circulation is good, you can be assured of impressive returns.

For their size, the amounts of produce that they give you are just phenomenal. However, they will also not overwhelm you with produce that is far too abundant for your use and consumption.

Basically, if you have a small family or are simply not that keen with a bumper crop of cucumber fruits, choose the bush type. It is space friendly quality and this makes it all the more ideal.

Vining cucumbers

Vining cucumbers, especially if given the license to roam, use up quite the amount of garden space. When they are trellised however, they not only make great use of available space, but they may also be used as landscaping screens. Such varieties like burpee hybrid, dasher 11, country fair 83, slice master, saladin, sweet success, sweet slice or slice nice are immensely popular choices with regard to vining cucumbers.

The advantages of vining cucumbers

Although vining cucumbers will often demand for more planting space compared with the bush kind, they have been around for a longer time. As such, they have been bred into a much wider range of sizes and shapes. There exists no match, at least with regard to the bushing varieties, for such types as the lemon cucumber or the multiple white-skinned varieties of cucumber out there.

The vining kind often produces amazing amounts of fruit, especially when you compare them to the bush variety. This leaves you with ample extras for trading to friends, picking, and the like.

*Key point/action step
Each type of cucumber requires a unique approach if you are to get the most output from your garden. You should therefore understand the benefits that come with growing each type of cucumber before you start.

Trouble Shooting, With Regard To Healthy Cucumbers, Harvesting & Storing Your Cucumbers For Best Preservation Of Quality

"Just knowing what the problem is is solving it halfway."

- Allen Ginsberg

Troubleshooting, with regard to healthy & tasty cucumbers

If you discover that your vines are blooming but there is a marked absence of fruit, chances are that there is something getting in the way of effective pollination. First of all, make sure that you observe both the male and the female blooms. The male blooms will be the earliest to appear, before dropping off. If this happens, there is no cause for alarm. Within a couple of weeks at most, female flowers will appear, each one with a small swelling at the base, shaped like a cucumber. This is what later develops into a cucumber.

There are several pests that are a bother to cucumbers. Squash bugs have a very defined affinity for cucumber seedlings. The slug family on the other hand waits until the fruit has formed and is ripening before moving in. The aphids love to colonize not just the leaves, but the buds as well. Straw mulch gives slugs a very uncomfortable time, so you should opt for it, as well as setting up trellises which lift the fruit off the ground.

Cucumber beetles also bother vines to a large degree, chewing holes in the leaves as well as the flowers and leaving deep scars in the stems and fruits. Worse than this however, they spread a disease that causes the cucumber plant to wilt off and die. Powdery mildew is a cucumber disease that leaves white patches on the cucumber leaves. At the first sign of its presence, comence on the application of fungicides.

So as to cut down any spread of disease, do not harvest your crop when the leaves and vines are still wet.

Harvesting and storage

Whenever your cucumbers are ripe and big enough for consumption, you may go about picking them.

Check the vines every day as the cucumber fruits start to appear because they tend to enlarge very quickly. The more you harvest, the more fruit the vines produce.

To remove the fruit cleanly, use some clippers or a knife, taking care to cut the stem just above the fruit. Tugging at the fruit will only leave you with a damaged vine.

Do not allow the cucumber fruits to get too big. The reason for this is that they become bitter tasting. They will also keep the vine from giving more produce.

Yellowing at the bottom of your cucumber is a signal for overripeness. You must remove the fruit immediately.

Harvest your lemon cucumbers just before they turn yellow. Although a major reason why they are called lemon cucumbers is because they turn yellow when ripening and end up looking a lot like a lemon (the shape contributes to this too), allowing the fruit to turn yellow may result in a taste that is a little too seedy for most people's tastes.

Keep those cucumbers that you harvest in the refrigerator for a period of 7 to 10 days. However, the recommendation is to use them as soon as you pick them. This is to get the flavor while it is at its best.

If you do not eat a slicing cucumber all in one sitting, wrap up what remains in plastic wrapping so as to prevent dehydration. Store in a refridgerator.

*Key point/action step
You need to follow specific techniques to determine if your cucumbers have the best yield. You will also need to be aware of the best time to harvest if you really want to get the tastiest cucumbers from your garden.

The Top 21 Must-Know Tips For Healthy & Tasty Cucumbers That Your Competition Is Probably Unfamiliar With

"Any extra thing that keeps you above your competition, no matter how small it is, embrace and own."

- Napoleon Hill

Nearly one half of the nation's veggie growers, roughly 47 percent for specifics, plant cucumbers according to one Susan Littlefield, the National Gardening Association Horticultural editor. This effectively makes cucumbers the number two most popular veggie that is homegrown. It should surprise no one that tomatoes are the top grown veggie, at a whopping 86 percent. Still, the cucumber percentage value is immensely impressive.

If you have a garden space that gets maximum sunshine, then growing your cucumbers will be an easy enough practice. Especially if you follow all of these unique directives in this chapter and are not a victim of late freezes in the spring, you should begin harvesting superbly sized and tasty cucumbers in no time.

Planning and preparation tips

Always go for the disease resistant types. This is for obvious reasons- these will more effectively combat diseases thus boosting your chances of having a superb yield at the end of the day.

Always go for a fertile space that gets lots of sunshine every day. Cucumbers will do poorly in infertile soils. They may not make it at all if they are grown in places that are cold and receive no sunshine.

In order to have an earlier harvest as well as to greatly minimize the prospect of insect damage to your seedlings, start several plants indoors in individual pots. Trays with individual compartments are also a great substitute.

Set up several trellises or even a fence if you go for the vining kind. Trellises made of wire are best, as they make it easier for tendrils to wrap themselves around as the plant grows.

Planting tips

Only sow your seeds in your outdoors garden after the danger of frost has well and truly passed away, and are sure that come what may, the soil will retain sufficient warmth for optimum growth. Cucumber plants are extremely susceptible to frost damage.

Make your second sowing some 5 weeks later to get a late summer or early fall harvest. This is a good way to manipulate harvest timelines.

When you are seeding in rows, keep them neat. Plant your seeds about 1 inch deep and some 6 inches apart.

When you are seeding in hills do this: plant four seeds in one-foot sized diameter circles. Set them some five or six feet apart.

Caring for the cucumbers

When your plants are about 3 to 4 inches tall, start thinning them. Of course, this will depend on the type that you are working with (either pickling or slicing).

When growing your plants in hills, this is how to thin them: thin your plants to the healthiest two plants, in the phenomenon of plants having two or three leaves.

You must keep your soils evenly moist. Why is this? The reason is to keep your cucumbers from becoming bitter in taste.

About 4 weeks after you plant them in the soil, go about side dressing them. For very plant, apply two ample handfuls of compost. For each plant, keep the compost bands narrow.

After applying the fertilizer, apply a thick layer of mulch. Mulch will help keep the soil moisture locked into the soil, thus greatly benefitting the cucumber plants.

Tips on controlling pests and diseases

Be very keen in monitoring your cucumbers as well as other veggies that are in close proximity to them for any buildup of insect pests.

It could be that the most effective way for the home gardener to control pests, with the example of the very destructive cucumber beetle high up the list, is to involve habits that shake up the life cycle of these insects as well as their habits. These will include you covering your young plants with some lightweight row covers up until flowering sets in. Also, exercise crop rotation- it is very effective in this.

If you do decide that pesticides and insecticides are the way to go, try to stick to the natural as much as you can. The less toxic the insecticide, the better it will be to use. Now, the only trouble you may perhaps face with this is that cucumber beetles are a hardy lot. There are not many effective natural, non-toxic insecticides available to deal with them.

You might have heard of Kaolin clay. If you have not, then this book will shed some light on it for you. Kaolin clay is perhaps the most effective natural insecticide choice to deal with cucumber beetles. It acts as a potent, long lasting repellent to them.

There is a big problem with the usage of broad-spectrum kind contact pesticides (these include malathion, cyhalothrin, permethrin, carbaryl and pyrethrin). The problem with these is that they not only kill the pests, they also kill off the beneficial predators as well as the parasites of the insect pests.

Insecticides are pretty much something you may have to resign yourself to using. So here is what to do with regard to them: read all the package labels keenly. For example, be familiar with what they advice about application and, say, harvesting. Is it a stipulation to wait for a few days after application before you harvest? These are the kinds of things you are looking to know.

You ought to consider capturing the pest that is ravaging your cucumbers. However, this is not to be used as a substitute for pesticides or natural control- it is far too impractical to work anyway. Rather, capture the insect or pest, lace it in a sealed bag, and then take it to the local garden center. Ask the staff to have a look at it and then give you any useful advice they have on the best control method in your particular area.

Harvesting your cucumber plants

Harvest your cucumbers once they hit slicing or pickling size. Actually, do your harvesting every two days to prevent the probability of cucumber fruits from achieving an excessively large size as well as keep the plant consistently productive.

*Key point/action step

Growing cucumbers entails different aspects that you need to master if you are to get the best output. Right from selecting the type of cucumber to grow through preparing the soil, taking care of the cucumbers as they grow, keeping off pests and harvesting, you must be careful if you really want to get the most output.

How To Apply What You Have Learned?

This book has pretty much laid everything out as clearly and as straightforwardly as is possible. Most of the advice offered here is direct and comes with its own set of whys. So how do you go about using this book? This is easy enough. Simply exercise a much more practical approach to your cucumber tending compared to a more theoretical one. The book is very practical, often giving specific measurements, instructions, and directions. Thus, it will be of little help to you if you do not pick your patch out and set about planting your cucumbers.

This book directs you on growing the healthiest and tastiest cucumbers. Every directive is set to help you get the best. For the best results, try to keep within the confines given in this book to the highest degree that you can. This way, you can enjoy a superb produce that surprises even you.

How To Grow Healthy & Tasty Tomatoes

It's every tomato grower's dream to harvest juicy, ripe and tasty tomatoes from the garden. Whether you are a home gardener or a professional tomato grower, you need to understand the secret behind tomato flavor. You might be confused on how to achieve the ultimate tomato sweetness you desire. May be you are almost giving up hope of ever eating/harvesting a self-grown mouth watering tomato or you are just a beginner with no clue on how to go about it. The good news is that after reading this book, any lost hope will be restored and if you are a beginner, you will kick off the process of growing sweet tasting tomatoes in an enjoyable and fruitful manner. This book will help you know the various factors that determine tomato flavor, the most favorable weather and soil conditions for growing healthy tomatoes and the types of tomatoes cultivars you should select to achieve the ultimate sweetness you are yearning for. You will also learn how to prepare the garden, maintain soil pH and tips for tendering your tomatoes until they are ready for harvest or ready to eat.

The Basics

"Research is what I'm doing when I don't know what I'm doing."

-Wernher Von Braun

Before we get to further details, let's have a look at some few basics you should know about tomatoes.

Tomatoes and their flavors

Tomatoes are vegetables (call them fruits if you like) that come in various shapes, sizes, flavors, and colors. When it comes to sweetness, we all have different taste buds, which means that we differ on tastes-what you consider sweet may not be sweet for another. That said, tomatoes come in different flavors; acidic, tart, sweet, or mild and there are those that are generally considered sweet by most people. Plant genetics and garden variables such as rainfall, temperature, sunlight, soil type and the garden location determine tomato flavor. Flavor is a balance of sugar and acidity along with the influence of certain elusive, unpredictable compounds for flavor and aroma that every tomato breeder is eager to grasp-this is all about nature and its wonders. Tomatoes high in sugars and low in acids are generally sweet. Those high in both sugars and acids are considered by most people as having a more balanced taste while tomatoes low in both sugars and acids have a bland taste. In addition, always check the plant description to ensure you select the desired sweetness.

Here are some ways you can select tomatoes based on their flavor

Size of fruit; you might have heard of the saying that 'good things come in small packages' and in the case of tomatoes, it can't be further from the truth. The small sized cherry and grape tomatoes have more sugar than the full sized ones and therefore are considered to be sweeter. Cherries such as sun gold are exceptionally loved by kids and even grownups as snacks due to their very sweet taste. However, some full-sized tomatoes such as lemon boy, bush goliath and black krim are also sweet and medium sized ones like 'early girl' are sweet for salads and sandwiches.

Color of fruit; the tomato color also determines their sugar and acid balance. For example, yellow or orange tomatoes taste milder and less acidic than red tomatoes. When it comes to black tomatoes, some are created from a mixture of red and green pigments, which make them have a complex flavor which is loved by some and not so much by others. It's not necessarily true that a yellow tomato is less acidic than a black or red tomato—it all depends on the levels of sugar-acid combination as well as other compounds which gives it a milder taste. You can experiment tomatoes of different colors to find your most preferred sweetness.

Foliage; It is also important to know that the more dense and healthy the foliage on a tomato plant, the more it captures sunlight which is converted to sugars and other flavorful components. Therefore, plants with many leaves like heirloom varieties which include; the black cherry, chocolate stripes, delicious red beefsteak and so on, are considered to be tastier than those with scarce leaf percentage like the hybrid ones found in the market. According to the father of mountain pride and other common tomato varieties, Dr Gardner, heirlooms are sweeter due to their very soft texture when ripe and that their cells rapture quite fast to release the juice and the flavor components in the cells. This is not the case with the grocery store tomatoes which are meant to withstand rough handling for shipping purposes. However, there are many hybrid home garden tomatoes that are tasty. Just make sure the tomato leaves are healthy.

Key point/action step

The taste that your tomatoes ultimately have is dependent on many factors such as acidity, sunlight, nutrient levels, and other factors. In choosing the preferred cultivar, you need to factor such issues like foliage, color of the fruit, and your preferred size of fruit.

Factors That Influence Tomato Growth

"What is the use of a fine house if you haven't got a tolerable planet to put it on?"

-George Carlin

Regardless of the tomato variety you grow, external factors such weather can make a huge difference on tomato health and flavor. For instance, an identical variety can taste better when planted in California than when grown in the Deep South where nights are longer. Whether you are growing tomato seeds in a nursery or using purchased plant seedlings, in-house or outside, you need to provide the right atmosphere for proper growth. Given the proper conditions, you can grow tomatoes almost everywhere except in extremely cold weather. There are genetic and external growth-influencing factors. Here are the external factors that influence the tomato health and flavor.

Soil

Soil tops the list when it comes to growing healthy tasty tomatoes. Without good soil, your tomatoes won't grow to maturity leave alone sweeten. So, you need to ensure the soil in your garden is rich in all the nutrients required for seed germination and growth of a tomato plant. Your soil should be well aerated, have the appropriate pH and enough water. You need to learn as much as you can about your soil and these lessons are in the next chapter of this book.

Moisture supply

You need to regularly supply your tomato with water for them to grow strong foliage and healthy fruits, and to avoid cracking caused by sudden changes in moisture levels of your soil. Too little or too much moisture inhibits growth of plants. Good soil moisture directly translates to better uptake of nutrients and efficient manure utilization. There is nothing you can do concerning excessive rain but when watering is under your control, watch out for any tendency to overwater your tomatoes. Soaking your garden dilutes tomato flavor. You can manage any water menace by getting some good drainage or irrigation; dig trenches or whatever else that keeps off excess water!

Temperature

When you hear the word temperature, think about heat intensity. Just like us, these fruits love a warm environment and need an average temperature of 65 degrees F. (18 C.) or more to ripen.

Therefore, make a point of waiting until frost threats pass before setting your tomatoes in the garden. The ideal temperatures for growing tasty tomatoes are 50s or 60s nighttime and 80s daytime. Higher temps during days and night will trouble your tomatoes fruit growing process and lower temps will reduce the plant's ability to create flavor compounds. If the heat is too high, your tomatoes will lose more water, won't breathe well and won't take in water and nutrients, plus worms and other microbes will be destroyed. Freezing temperatures will kill your plant. This does not mean that if you do not have ideal temperatures, you can't grow flavorful tomatoes; just make sure you choose tomato varieties suited to your region. Heat is Key to getting a tasty tomato; there is a remarkable dissimilarity between a tomato that ripens in cool conditions and one that enjoys the benefit of good, hot summer days. Check with your extension officer to get advice on appropriate variety.

Sunlight

Quality, intensity, and duration of light are the most important aspects when it comes to sunlight. Tomatoes need areas with full sun and much protected from strong winds as well. Natural light is best for healthy leaf formation and fruit flavor. The sun's brightness takes full advantage of photosynthesis in tomatoes, letting the plants make carbohydrates that are eventually converted to flavor components—acids, sugars and other substances in the fruit. Giving your tomatoes 6-8 hours of intense sunlight daily greatly favors them so plant your tomatoes in a place that has enough access to light. Cloudy, wet regions without clearly defined day and night temperatures, like the Northwest, fail to produce the finest-flavorful tomatoes. However, heirloom varieties such as Seattle's and San Francisco Fog are known to perform better than most of other varieties in such areas.

Air composition

Grow your tomatoes in well-ventilated places. Co2 is converted to organic matter during photosynthesis and it is released to the atmosphere after that. Sufficient air circulation ensures that your tomatoes grow healthy and tasty. You should avoid air pollutants like the excessive chemical sprays which are toxic and suffocating to your tomatoes. Use ceiling fans to improve air circulation if you are planting indoors.

Biotic factors

Like any other plant, tomatoes are vulnerable to attacks from pests and diseases. Using fertilizers in excess makes plants even more susceptible to diseases so watch out on that. Weeds compete with your tomatoes for moisture, light, and nutrients so you should get rid of any weeds in your garden.

Plant nutrients

Similar to humans, for good health, tomatoes require the right balance of nutrients. For instance, if your soil is calcium deficient, your tomatoes will suffer from blossom-end rot. On the other hand, too much nitrogen accelerates leaf growth but can lessen the number of fruits or flowers. Nitrogen boosts the health of your tomato leaves, which adds on flavor. Yellow leaves indicate nitrogen deficiency. To increase on nitrogen levels, add organic manure, which is a healthier option. Studies have proven that the inorganic manures are full of harmful synthetic chemicals, so make a healthier choice. Organic sources include; alfalfa, fish meal, compost, leaf mold and feather meal.

Potassium helps in keeping diseases at bay and promotes tomato growth. Its deficiency slows growth and weakens the tomato plant. To boost on potassium level, use available organic substances such as granite dust, wood ash, and rock sands.

Phosphorus aids in formation of tomatoes' roots and seeds. Insufficient phosphorus in your soil can cause tomatoes to have reddened stems and stunted growth. If your test results indicate that your soil needs more phosphorus, you can add some bone meal or compost manure to boost it.

Key point/action step

Grow your tomatoes within the right environment for them to grow into healthy flavorful fruits, watch out for any changes in your plants' environment, and correct any negative ones if possible to avoid surprises in the outcome.

Soil Testing Strategies

"Studying wine taught me that there was a big difference between soil and dirt -dirt is to soul what zombies are to humans. Soil is full of life while dirt is devoid of it."

-Olivier Magny

Soil testing will help you know what needs to be done to make the soil ideal for tomato growth. Learning about your soil's acidity texture, drainage, composition, and mineral density will help you curb the frustrations that you may experience when your soil is unsuitable for a tomato garden of your dreams. You will get invaluable tips on how to do soil testing in this chapter and for sure, you will see that soil testing is not rocket science.

First, you need to prepare a soil sample to use in the testing process. You can use the sample collected to do the testing on your own or send to a soil lab if you can't do it for some good reason.

General guidelines for collecting a sample

1) Fill a cup with your vegetable garden top soil (4 to 6 inches from the surface) then put the soil in a plastic bag.

2) Dig soil samples from different parts of your plot. Obtain six to eight similar samples then put them in the plastic bag.

3) Combine well the soil from all the cups; put two cups of the mixed soil in a different plastic bag —you have your soil sample!

After you've collected your sample, you can take it to the lab or do the testing yourself to get more hands on experience and understand your soil better.

Here are several soil tests you can do on your own:

Soil Test#1: The Squeeze Test

Soil composition is one of the most basic characteristic. Soil is broadly classified into 3: clay, loam, and sandy soil. Clay is slow draining but rich in nutrients, sand is fast draining but doesn't retain nutrients while loam is the considered the mother of all soils and ideal for planting delicious tomatoes and almost all other crops. Loam is rich in nutrients and retains moisture without being soggy.

Steps to test your soil type:

1) Take a handful of moist soil (not wet) from your garden.

2) Squeeze it firmly then release your hand.

3) Stroke it lightly. If it retains its shape but crumbles when you stroke it, you are very lucky to have the luxurious loam in your garden.

Soil Test#2: Soil Drainage Test

Testing your soil's drainage is equally important when it comes to planting sweet tomatoes. A waterlogged garden makes tomatoes tasteless, which is why if you pick a tomato the morning after it has rained, you will notice it's not as sweet as it was before the rain.

Steps for checking soil drainage:

1) Dig a hole; one foot deep and six inches wide.

2) Pour water into the hole up to the brim and let it drain off completely.

3) Fill it with water one more time.

4) Record the time it takes for the water to drain each time.

If it takes more than four hours for the water to drain, you have poor soil drainage and you need to improve it by digging trenches or applying other methods available for soil drainage problem; tips are available online or inquire from extension officers.

Soil Test#3; The Worm Test

Worms are wonderful indicators of how healthy your soil is, in biological activity terms. If you see earthworms coiled up or moving in your garden, you should rejoice because their presence means that there is a high chance that all the bacteria, nutrients, and microbes necessary for a healthy soil and strong tomato plants are present in your garden. A dead soil destroys all life forms!

Steps for checking on worms:

1) Ensure your soil is at least 55 degrees warm and somehow moist, although not soaking wet.

2) Dig a hole one foot deep and one foot wide. Put the soil on a cardboard piece or a trap.

3) Sift the soil through your hand as you put it back into the hole and count your blessings-earthworms. Don't fear the worms; they don't bite!

If you have at least 10 or more worms in your hand, your soil is in excellent shape. Less than 10 worms is a red flag for insufficient organic matter in your soil to support worm population growth or that your soil is too alkaline or acidic.

Soil Test#4; Nutrient And pH Test

Soil pH (acidity) and nutrients has a lot do with how well your tomato plants grow. Soil pH level is rated on a scale of 1 -most acidic and 14-most alkaline; at level 7, the pH is neutral. Levels of below 5 or above 8 will stunt the growth of your plant. Tomatoes require a slightly acidic soil with about 6 to 7 pH level. You can do the acidity test on your own using the approved soil test kits available in local garden stores or online. The kit measures the acidity and the nutrient content in your soil. To get accurate results, follow the instructions on the kit to the letter.

In case you have tested and amended your soil but still experiencing recurring problems with the soil, contact your local agricultural extension officer who will test the soil in a lab and alert you on your soil's mineral deficiency and how to solve any underlying issues.

The above tests are inexpensive and simple ways to guarantee your garden has the best foundation possible for growing your tomatoes. Once you have performed all these soil tests and ascertained that your soil is at the best condition possible, it's time for you to get down into the business of preparing your garden for planting your tomatoes.

Key point/action step

Test your soil's acidity, nutrient level, composition, drainage and for worms. Without this, you can't really understand your soil condition so go ahead and do the testing now. Get the do-it-yourself kits!

Prepare Your Soil And Plant Your Tomatoes

"The soil is the great connector of lives, the source and destination of all. It is the healer and the restorer and the resurrector, by which disease passes into health, age into youth, and death into life. Without proper care for it, we can have no community, because without proper care for it, we can have no life."

-Wendell berry

Soil preparation tips

Soil preparation is a crucial step you should take before you start to plant your tomatoes. A well-prepared soil produces first-class tasty tomatoes. When preparing your soil, keep in mind that chemical fertilizers may promote growth of the plant, but do not increase on the flavor of the fruit and in fact can make tomatoes hardened and bland, use organic manure for healthy and tasty tomatoes.

Here are simple guidelines of preparing the soil in your garden:

Warm the soil

Start the soil preparation by warming the soil on which you plan to grow your tomatoes. Tomatoes do well in warm soil. Add gravel to the soil, which helps with drainage and raises the soil temperature. Either you can wait for the air temperature to rise, which will take a while or you can simply cover your soil with a black plastic paper to help in moisture absorption. You can use bricks, rocks or anything else sturdy and heavy to secure the plastic firmly on the ground just in case strong wind blows.

Test your soil's pH level

Use the soil acidity test kits as mentioned earlier. If needed, you can adjust your soil's pH level. If the pH is too high, put some sulfur in it and if it's too low, put in lime. If your soil acidity is not suitable, your tomato plant can't absorb all the nutrients required for proper growth, even if your soil has them in large amounts. If the acidity is too low, it will increase solubility of minerals such as manganese making your tomatoes toxic.

Evaluate your soil's nutrient level

Use the acidity test kit or take a soil sample to a local approved lab for testing. The test will show you the chemical makeup of your soil and the nutrients in your soil. There should be a good balance of potassium, nitrogen, and phosphorus in your soil for you to yield good tasty tomatoes.

Add compost

A great way to improve your gardening soil is put some compost to help improve the soil cultivability, structure and nutrition retention. It also attracts earthworms and increases microbes. Compost is made up of broken down organic matter. You can purchase compost manure in a gardening store or you can make your own using leaves, fruit and vegetable wastes or yard clippings. Add plenty of manure-based or spent mushroom compost to your soil. Dig a roomy hole and mix the soil with the compost. Whether using a pot or your garden, work a half-inch of compost into the soil.

Once your soil is setup, you can now go ahead and plant your seedling indoors or outdoors.

Planting process

You can start to grow your tomatoes from seeds, which will not only offer you a variety of choices, but also costs less. If you are using seeds from a ripe tomato you just ate, make sure they are dry and fermented and from a good plant like the heirloom or they are open pollinated seeds.

1) Put the seeds in a container with water and place a loose fitting lid on top to allow oxygen to enter. Label the container to avoid mix-ups.

2) Put the loaded container in a warm place somewhere far from you to avoid the awful smell. Wait for 2-3 days and stir the mixture daily until you see some molds on the surface, then remove the mold with some gloves on.

3) Pour some more water into the container to dilute the mixture, pour out the unwanted solution then sieve out the seeds and rinse them before they germinate. Dry your seeds on a non-stick surface like a baking sheet for several days. Store in sealed plastic bags or in the fridge (not the freezer) inside airtight containers for later use. Label the containers and bags!

If you buy them from a garden or a nursery, make sure you select bushy plants without flowers and lookout for presence of any pests.

Steps for planting seeds

1) Sow your tomato seeds indoors before taking them outdoors in pots or trays for around six weeks before the expected end of spring frost in order to avoid stunted growth or even death of your plant. The proper pots (peat pots or other small pots) are available in garden stores or local nursery.

2) Make sure you fill your pot with soil mix, for example 1/3 course vermiculite, 1/3 peat moss and compost. Just find a good mix online or seek extension services.

3) Sow the seeds in holes 2 to 3 inches deep inside the pot or tray. Sow twice as many seeds so that you can be able to select the healthiest and strongest seedlings to grow in your garden.

3) Store the containers in 70 to 80°F (21-27°C) room temperature. Germinating seeds require a temperature of around 75-89°F but in reality, the seeds can germinate within normal indoor temperatures of about 68-73F. To get more heat, you can put the seeds on your fridge. After they germinate, place them where they can get sufficient direct light from the sun or near the window during winter. If push comes to shove, you can use grow lights -not the so expensive ones but good bright white lights. They are not as strong as the sun so place them close to the plant as much as possible but not too close to burn the plant.

4) Mist tomato seeds daily for the initial 7-10 days. When the first sprouts appear, water less frequently. Keep checking your pots daily for plants peeking out of the soil.

5) Cut off the plants you don't need with scissors to avoid space and nutrient wastage; if you planted twice as much.

6) Transplant your seeds to 4 small pots if you are sowing them in a tray, because the roots will run out of room to grow. After the first true leaf appears, hold it carefully using your thumb and fore finger with one hand and use a chopstick, pencil or any similar object to

dig into the soil and loosen the roots (don't hold the stem). Put them in a pot with well-prepared soil. Put the seedlings at the centre of the pot and pour dirt over the roots, avoid pressing the soil down; you might damage the roots so just water them and everything will fall into place.

7) Take your plants outside regularly a week before transplanting them to your garden, under bright light of 6-8 hours in order to harden them off. After your plants are above 2 inches, you need to place support in order to help them grow strong.

8) Transfer your plant outdoors. Once your nighttime temperature is consistently higher than 50 degrees and your tomatoes are 6 inches (15.2 cm) tall, you can transfer your plant to your well prepared garden. Dig a hole of approximately 2 feet deep and put in some organic manure. Take off some lower leaves to make sure you plant them deep enough. This will promote root growth, which translates to better uptake of water and minerals, and avoid plant water loss. Plant the tomatoes in simple rows and use about 8-10 seedlings in each row for a small, manageable garden. Add peat moss to your soil to improve its drainage if you desire or build a raised garden using a good wood like cedar.

Key point/action step

Set your soil up for planting by warming your soil, balancing nutrients and acidity levels, add organic and compost manure to ensure proper growth, health, and flavor. Plant your seedlings in this loaded soil and lovely environment. Put this in practice and you won't be disappointed by your results!

How To Care For Your Growing Tomatoes

"A garden requires patient labor and attention. Plants do not grow merely to satisfy ambitions or to fulfill good intentions. They thrive because someone extended effort on them."

-Liberty Hyde Bailey

Tomato plants require extra care to ensure that they grow into healthy and flavorful fruits. We can't talk about caring for healthy tomatoes without touching on pests and diseases which greatly affect tomato yields, health and flavor. Presence of diseases and pests suggests that something is not right in your plant internal and external environment. Plant diseases and pests are mostly due to poor conditions such as inadequate water, nutrients, space or sun; pathogens such as bacteria, fungi, or viruses; and weather. However, with proper maintenance and care, you can overcome most of these problems easily. Just like humans, you need to boost your plant's immunity. Also, if your area is prone to certain types of diseases or pests, make sure you choose the kinds of tomatoes that are listed as resistant. Try to grow three to four tomato varieties to see which one suits your locality, which is susceptible to diseases and which one tastes better. If growing tomatoes outside, you can start with cherries or black krim for they do well in most areas and ripen faster than others do.

Here is how to care for your plants and keep off the pests and diseases:

Mulch your tomatoes

Once the soil is fully warmed up, you can do mulching which helps to suppress weeds, retain moisture, and avoid disease problems. Don't mulch too early to avoid prolonged cool underground temperatures. In the fall, you can plant living mulch called hairy vetch. You can mow it down in spring, and plant tomatoes through it, which works very well. Several studies propose that hairy vetch mulch enhances the tomato plant's ability to utilize nitrogen and calcium and increases their resistance to disease. Other mulches, such as chip mulch, wheat straw, promote the plants' roots and prevent rain from splashing soil-borne disease microorganisms onto the foliage.

Keep your plants upright

Maintain your plants in an upright position, by growing them in sturdy cages or secure them to trellis or stakes-this largely depends on your space. This keeps the foliage high over the ground, which increases the chances of each leaf's exposure to the sun and lessens the risk of foliage loss due to disease in addition to making fruit picking easier. Just make sure you don't destroy the roots in the process.

Water your plants

Never let your plants droop due to insufficient water supply. You need to water your plants if the weather is warm and dry. Water the plant deeply in the morning hours for approximately once to thrice a week. Avoid watering at night, which endangers your plant, as insects prefer wet dark environments and it makes your tomato vulnerable to diseases such as rots and mold. If you water during noon, the water will evaporate very fast, even before your plants absorb. You can sink in a pipe vertically on the ground when you plant the seedling on the garden, to make sure the water gets to deepest roots faster. Water your plant at the ground level and not on the leaves to prevent diseases. Water the soil not the plant stems or leaves!

Feed your plants

Just keep the nutrients coming. Every week after your plant starts flowering, give them a comfrey or seaweed feed to increase fruit production. Fertilize your tomatoes immediately after planting using premixed high phosphorus-low nitrogen organic manure to avoid diseases, promote growth and taste. Do this to your garden once a year.

Remove the plant suckers

You can cut off any shoots that form between the main branches and the stem as your plant grows-they just consume valuable energy from the emerging fruit. Just leave a few stems near the top to prevent sunscald. When growing tomatoes outdoors, cut off the tops once the first six fruit trusses appear in order to focus the plants energies.

Harvest the fruit at peak time

Tomato fruit should emerge six days after transplanting. Keep an eye on the plants daily once they start ripening, to ensure you get maximum flavor. Once the fruits are fully ripe, pick them by gently twisting the fruits and not pulling out the vine.

Observe these special warnings

Never put tomato seeds in direct sun under temperatures above 85° to avoid damaging your seeds. Common tomato diseases are fusarium and verticillium wilt, which you can prevent by planting resistant cultivars/types, crop rotation and good hygiene. Common pests that may plague your garden are white flies, nematodes, and cutworms, which, you can also control by maintaining the proper plant environment and use of manure. If problems persist, seek your extension officer's advice.

Key point/action step

Mulch, water, prune the extras and straighten your tomato plant in order for you to have a good, healthy and flavourful harvest. Just make sure the fruits are glowing in ripeness before you pluck them off the vine.

How to Apply What You've Learned?

Growing healthy and tasty tomatoes is easy once you've identified the ideal cultivar for your area depending on the weather, temperatures, soil and other aspects. And even if some of the conditions in your area are not ideal for the particular cultivar that you want to grow, you can always adjust such conditions as soil pH, nutrients, and drainage to ensure that you only provide the very best conditions for your tomatoes to thrive.

It is best to plant tomato seeds indoors then move them outside when they have a few leaves after about 6 weeks. Once you've transplanted them, you will then need to watch out for pests and diseases, protect them against harsh weather conditions, provide sufficient sunlight, monitor the temperatures, water them properly and ensure that they are fed properly if you want to harvest tasty tomatoes in the end.

www.ingramcontent.com/pod-product-compliance
Lightning Source LLC
Chambersburg PA
CBHW071827200526
45169CB00018B/1104